Advanced Analysis in Steel Frame Design

Guidelines for Direct Second-Order Inelastic Analysis

SPONSORED BY
Special Project Committee on Advanced Analysis of the Technical
Committee on Structural Members

The Structural Engineering Institute (SEI)
of the American Society of Civil Engineers

EDITED BY
Andrea E. Surovek, Ph.D., P.E.

Published by the American Society of Civil Engineers

Cataloging-in-Publication Data on file with the Library of Congress.

American Society of Civil Engineers
1801 Alexander Bell Drive
Reston, Virginia, 20191-4400

www.pubs.asce.org

Preface

The idea for a set of published formalized guidelines for the use of second-order inelastic, or "Advanced", analysis began in Structural Stability Research Council (SSRC) Task Group 29, Second-order Inelastic Analysis for Frame Design, chaired by Donald W. White. The task group published the report "Plastic Hinge Based Methods for Advanced Analysis and Design of Steel Frames" in 1993. The current effort is aimed at bringing the knowledge gained from the previous report and subsequent research efforts into a coherent set of guidelines that places Advanced Analysis in the context of the current AISC Specification while being general enough to adapt to other standards and to future editions of the AISC Specification. The guidelines are designed to provide a methodology for Advanced Analysis that gives the designer specific recommendations on rigor of analysis, minimum modeling requirements, consideration of limit states, serviceability and live load reduction, while allowing latitude for the judgment of the design engineer.

Members of the Committee involved in the development of the guidelines include:

Andrea E. Surovek, Chair	South Dakota School of Mines and Technology
Bulent Alemdar	Bentley Systems, Inc.
Dinar R.Z. Camotim	ICIST/IST Technical University of Lisbon
Jerome F. Hajjar	Northeastern University
Lip Teh	University of Wollongong
Donald W. White	Georgia Institute of Technology
Ronald D. Ziemian	Bucknell University

ACKNOWLEDGEMENTS

The development of this report was supported by a special project grant of the Structural Engineering Institute of ASCE, and was completed under the guidance of the SEI Metals Technical Committee on Structural Members (formerly the Committee on Compression and Flexural Members). The research contributions of past and current members of the SSRC former Task Group 29 and current Task Group 4 were instrumental in developing this report.

iii

Contents

PART 1:
INTRODUCTION

This document presents recommendations for the use of second-order, inelastic analysis in the design and assessment of steel framing systems. These guidelines can aid engineers, specification developers and structural analysis software developers in understanding the baseline requirements for directly capturing member and system strength limit states when using a second-order inelastic analysis. In traditional design, the load capacity of the system is assessed on a member-by-member basis, limiting the load carrying capacity of the system to the strength of the weakest member. Alternatively, an "advanced analysis" methodology focuses on the structural *system* strength rather than limiting the strength of the structural system at design load levels by the first member failure. The guidelines include analysis and modeling requirements as well as design considerations (e.g. suggested resistance factors) such that the behavior and strength of the overall system and the limit states of individual members are checked concurrently without the need for separate in-plane specification member strength checks.

1.1   What is Advanced Analysis?

The design of metal framing systems in most international design standards requires the use of second-order forces when strength checks are performed. These forces are typically obtained directly from one of the following types of elastic analyses:

- a second-order analysis that explicitly includes second-order effects in the analysis algorithm and requires an incremental iterative solution

- an approximate second-order analysis in which second-order effects are approximated by some means within the context of a linear analysis (e.g. P-Δ analyses approaches); or

- a first-order elastic analysis in which the forces obtained from the analysis are multiplied by amplification factors to obtain approximate second-order results.

The second-order forces and moments are then used in member strength checks. It is important to note that the capacity of the *system* is never directly checked in this approach, but rather each member is individually checked to determine if the capacity of that member had been exceeded. In other words, the capacity of the structural system is essentially bounded by the first member to reach its maximum capacity or fail a strength check. Because the strength of the system as a whole is never directly assessed, the ability of the system to inelastically redistribute loads at maximum design load levels is generally not considered. In addition, the overall structural performance is not directly assessed in methods that focus on individual component strength rather than focusing on system behavior. While some approximate system-based approaches may be employed in design of smaller structures, such as plastic

1

mechanism checks, the stability effects on the full inelastic strength of large (e.g., highly redundant) structures can only be assessed rigorously through the use of a nonlinear, inelastic analysis approach.

While current specifications rely predominantly on the elastic second-order analysis results described above, a wealth of literature has been produced in the last 25 years on capturing the behavior of steel framing systems through the use of robust nonlinear, inelastic analysis. The use of a nonlinear, inelastic analysis to directly determine the overall strength and stability of a steel framing system is commonly referred to in the literature as "Advanced Analysis" and is defined as "any analysis that accurately represents the behavioral effects associated with member and system primary limit states to the extent that corresponding specification checks are superseded" (White and Chen 1993). This is accomplished by incorporating the fundamental attributes and behavior associated with most member limit states directly in the analysis. When these effects are captured directly in the analysis, separate member checks are not required for those limit states. *Any limit states that are not directly modeled must still be accounted for through separate member or component strength checks.* Depending on the complexity of the member elements used in the analysis, these limit states may include out-of-plane, torsional or local buckling effects. Additionally, limit states related to rupture are not directly captured.

Decades of research have produced significant advances in capabilities for efficiently capturing frame response using second-order inelastic analysis. In addition, modeling requirements, serviceability considerations and design procedures have been studied when the strength of structures is explicitly assessed in the analysis. Papers and reports on these topics include, but are by no means limited to the following: Ziemian 1990, Clarke et al 1992, White 1993, White and Chen 1993, Chen and Toma 1994, Maleck et al 1995, McGuire 1995a & b, Chen and Kim 1997, White and Nukala 1997, Alemdar 2001, Maleck 2001, Deierlein et al 2002, Trahair and Chan 2003, Alemdar and White 2005, Surovek and Ziemian 2005, Martinez-Garcia and Ziemian 2006 and White et al. 2006. *This report addresses the synthesis of these many efforts into a practical design methodology for assessing the strength and behavior of steel framing systems using advanced analysis procedures.*

The 2005 AISC specification (AISC 2005) marked a move forward in allowing more direct assessment of steel framing systems with the inclusion of Appendices 1 and 7, Design by Inelastic Analysis and the Direct Analysis Method (DM), respectively. In the most recent AISC Specification (AISC 2010), DM has been moved into Chapter C as the preferred method of frame stability assessment, and the traditionally used effective length approach has been moved to Appendix 7. One of the key components of both the DM and design by inelastic analysis provisions is that they focus on obtaining more realistic analysis results. By incorporating a nominal out-of-plumbness and a nominal stiffness reduction in the structural analysis, the DM permits the checking of steel frame strength accounting explicitly for the key phenomena that affect the system and member strengths (Maleck 2001, Deierlein 2003, Surovek-Maleck and White 2003, White and Griffis 2007, Griffis and White

2007, Kaehler et al. 2007). Major advantages of the DM are: (1) it eliminates the need for effective length factors, (2) it provides an improved representation of the internal forces throughout the structure at the ultimate strength limit state and (3) it applies in a logical and consistent fashion for all types of frames including braced frames, moment frames and combined framing systems (White and Griffis 2007). In the DM, the effects of inelasticity and frame imperfections are accounted for in the calculated forces required for the design of all the structural components. One of the benefits of this approach is that it provides a direct path from elastic analysis and design to advanced analysis and design (Surovek and Ziemian 2005; White et al. 2006).

Appendix 1 of the 2010 specification states that "strength limit states detected by an inelastic analysis that incorporates [the following] are not subject to the corresponding provisions of the Specification when a comparable or higher level of reliability is provided in the analysis." The following attributes are required in the analysis:

> "(1) Flexural, shear and axial member deformations and all other component and connection deformations that contribute to the displacements of the structure; (2) second-order effects (including P-Δ and P-δ effects); (3) geometric imperfections; (4) stiffness reductions due to inelasticity, including the effect of residual stresses and partial yielding of the cross-section; and (5) uncertainty in the system, member and connection strength and stiffness." (AISC 2010)

Appendix 1 is not prescriptive in how these attributes are to be included in an inelastic analysis. One of the objectives of this document is to provide guidelines for how the requirements of Appendix 1 might be met while also providing the necessary background to understand the rationale for the recommendations.

1.2 Motivation and Scope

Stated succinctly, advanced analysis simplifies the structural strength assessment and gives the engineer greater design flexibility.

Advanced analysis capabilities, if properly implemented and applied, simplify the checking of the strength limit states. If an advanced analysis shows that the structure is able to support the required factored loadings, certain specification-based primary member strength equations are inherently satisfied. Indeed, separate checks of the corresponding column, beam and beam-column resistance equations are, for the most part, no longer necessary. While greater complexity is introduced in the analysis, a significant reduction in effort is achieved in the design assessment. Conventional wisdom suggests that the design of many types of structures is governed by serviceability rather than strength. Advanced analysis is particularly beneficial in these cases since the primary goal is a verification of the overall strength. This may be accomplished through an efficient final checking of both member and system limit states for a structure where preliminary member sizing was based on serviceability requirements.

However, perhaps the greatest benefit of advanced analysis is that it facilitates a holistic approach to the design of framing systems can be used as a tool to explicitly meet performance-based design objectives. Because the member and system strength may be directly assessed, including the inelastic reserve strength, the designer is afforded a greater freedom in tuning the structural design to achieve design objectives. For some types of structures, this can avoid problems such as those that may occur when using the effective length method, where design modifications increase the effective lengths and the process leads to larger member strength unity checks when the member sizes are increased, (e.g., see Rex and Goverdhan 1998 & 2002).

Another strong rationale for advanced analysis is that it provides more realistic values for internal moments and forces. Serviceability limit states such as limits on permanent (inelastic) deformations or total deflections at desired service load levels also may be checked most rationally by using advanced analysis, since this type of analysis characterizes the actual system behavior more accurately than elastic analysis approaches.

Advanced analysis is geared towards the engineer who understands system behavior and wishes to both expedite design checks and have greater flexibility in design. This report presents various criteria that should be met when taking advantage of these analysis tools, along with recommended ways to satisfy these criteria. In particular, minimum standards are presented that must be met for an analysis approach to be considered an "advanced" analysis.

1.3   Basic Requirements for the Guidelines

These guidelines are based largely on SSRC Memorandum 5 (Galambos 1998) "General Principles for the Stability Design of Metal Structures", which states the following with respect to assessment of the strength of framing systems:

1. *Whenever possible, the procedure for the establishment of the load carrying capacity of frames, members or elements on the basis of maximum strength should be based on a mathematical model that incorporates:*
2. *Experimentally determined physical characteristics, such as residual stresses, material nonlinearities, and cross-sectional variations in yield strength, rationalized as may be appropriate.*
3. *A statistically appropriate combination of acceptable characteristics that are specified in supply, fabrication, and erection standards, such as out-of-straightness, underrun of cross-section, cross-sectional dimensional variations, material properties and erection tolerances.*
4. *Effect of boundary conditions, such as restraint applied to the end of members.*

These principles are incorporated in the general requirements that Section C1.1 of the 2005 AISC Specification imposes on all analysis and design methods as well as the requirements of Appendix 1. There are three major considerations necessary to meet the SSRC Memorandum 5 requirements, as well as the AISC (2010). These include:

1. Level of analysis (required level of rigor)
2. Aspects to be included in the physical modeling of frames
3. Design considerations

The first two relate to accurately capturing the behavior and maximum strength of the structure based on the aforementioned phenomena. The third places advanced analysis in the context of design Specifications and Codes and design office practices, including recommended handling of resistance factors, application of live load reduction and consideration of serviceability.

1.4 Limitations of the Guidelines

This set of guidelines addresses analysis and modeling requirements for steel frames subjected primarily to bending in one plane.

At present, there are analytical approaches that capture three-dimensional limit states for beam column behavior (e.g. see White and Nukala 1997). However, these are not commonly employed in academic software packages, much less in commercial ones. Consequently, it is important to note that the provisions are primarily applicable to planar frames, or more specifically, to structures in which beam-column members fail in the plane of bending. While this does not preclude the recommendations contained in this document to be applied to three-dimensional framing systems, it does require a higher degree of prudence on the part of the engineer to ensure that potential limit states due to three-dimensional effects, such as flexural-torsional buckling, or lateral instability of beams, are adequately captured by checks outside of the analysis.

Section H1.3 of the 2010 AISC Specification provides equations for separate checking of the out-of-plane strength of doubly-symmetric members. These provisions allow out-of-plane moments to be neglected when the strength ratio M_r/M_c in the out-of-plane direction is smaller than 0.05 (this may be considered as a rational limit for general inelastic analysis and design). Section H1.3 may be used to check member out-of-plane resistances in cases where it is applicable. Alternatively, Section H1.2 may be used to check member out-of-plane resistances. In cases where members and other structural components are required to withstand significant inelastic deformations, it is assumed that the members and/or components are designed to ensure that their inelastic deformation capacities are greater than or equal to the inelastic deformation demands. The provisions in Appendix 1 of the 2005 AISC Specification give one accepted way of satisfying this requirement. Members and components that do not satisfy the requirements of Appendix 1 may be designed elastically based on forces determined from an advanced analysis. The guidelines are applicable to braced frames, moment frames and combined systems in these contexts.

The extension of these provisions to three-dimensional structures is reasonable as long as any limit state not considered in the model is considered separately. In addition, three-dimensional effects not specifically addressed by this document (the most notable being torsional effects) must be addressed by the engineer in a rationale fashion, either based on current research methods or separate member or system checks. For example, if torsional flexural response is significant in the strength of the member, this must be separately checked, and any coupled torsional effects that impact inelastic response must be explicitly considered.

It is important to note that the use of an advanced analysis approach substantially reduces the required component checks, but it rarely eliminates them completely. However, it does still provide a more comprehensive picture of the overall system behavior up to system failure.

It is also worth noting that it is not possible to exactly model everything that might contribute to the overall behavior of the structure; consequently, some degree of approximation is introduced any time a simplification of the model is employed (e.g. excluding finite sizes of connections or panel zones). In instances where modeling may be simplified, the onus is on the designer to determine the acceptability of the effects of the approximation on the overall result. However, the engineer is cautioned against compounding approximation errors when using multiple simplifications in the model.

PART 2:
DESIGN RULES

The following rules present the minimum requirements for the analysis, modeling and design using an advanced analysis approach.

2.1 Analysis Requirements

2.1.1 Rigor of analysis required

The analysis model must be able to represent the reductions in member stiffness due to:

- The spread of plasticity through the member cross-sections and along the member lengths
- In-plane stability effects of axial forces acting on the inelastic member deflected geometry.

The model should accurately represent the reduction in member stiffness when compared to the benchmark solutions in Section 4. The satisfaction of these benchmarking requirements ensures that the in-plane strength limit states are captured comprehensively in the second-order inelastic analysis.

2.1.2 Acceptable analysis approaches

Various methods of second-order inelastic analysis have been proposed in the literature that can potentially satisfy the requirements of Section 2.1.1. This report recommends two types of analysis approaches for advanced analysis:

- Distributed plasticity analysis
- Refined plastic hinge analysis

The specific attributes of these approaches are discussed in Section 3.1

2.2 Modeling

As outlined in Section 1.3, all analysis and design procedures must account for the attributes that significantly influence member and system strength. The attributes specifically addressed in advanced analysis models are outlined below.

2.2.1 Inelasticity

Inelastic material behavior must be included in the analysis model. This may be accomplished by using one of the analysis approaches recommended in Section 2.1.2.

2.2.1.1 Residual stresses

The member analysis formulation or constitutive model must incorporate the effects of residual stresses in rolled or built-up members. The residual stress pattern chosen should be appropriate for the type of cross-section being considered.

7

2.2.2   Geometric imperfections

Member and system imperfections may be classified as out-of-plumbness (i.e. frame nonverticality) or member out-of-straightness (member sweep). Three-dimensional effects, such as member out-of-plane and/or torsional imperfections, are not addressed in these guidelines. Effects of imperfections should be included for all load cases, as specified below.

2.2.2.1   Out-of-plumbness

General Requirements: It is preferable to model out-of-plumbness, or frame nonverticality, by modification of the frame geometry; however, in orthogonal frames, imperfections may be modeled through the use of equivalent horizontal notional loads, proportional to the gravity load, applied at each story level. The out-of-plumbness should be modeled to a degree that accurately captures the potential nonverticality that can occur during construction. It should be modeled in the direction that is most detrimental to the strength of the system at the design load level; typically, this will be the direction in which the frame sways under the applied loads. In cases where symmetrical structures are loaded symmetrically, the symmetry of the design should be maintained in the component selection.

Recommended Approach: For rectangular frames less than 85 feet in height, model the out-of-plumbness by modifying the vertical geometry by H/500 over the height of the frame, where H is the frame height. For frames taller than 85 feet, one of two methods may be used:

- Model a linear out-of-plumbness of H/500 over the height of the building in 85 foot intervals. This requires analyzing multiple models such that imperfections throughout the entire height of the building are considered.
- Using a linear variation over the height, model the maximum imperfection based on the AISC erection tolerances (2" at 1000" of height plus 1/16" per story up to 3" thereafter) at the top story.

For gabled frames, shift the nodes horizontally by H/500 in both the columns and the rafters in the orientation that is most detrimental to the stability of the frame, where H is measured relative to the lowest base elevation of the frame.

2.2.2.2   Out-of-straightness

General Requirements: Member out-of-straightness should be modeled in such a way that it represents the potential maximum destabilizing effects due to a corresponding physical member imperfection. It only needs to be modeled explicitly in an advanced analysis when it has a significant effect on frame behavior. In general, if including the imperfection decreases the member capacity by less than 5%, the imperfection may be neglected in the analysis.

Recommended Approach: It is not necessary to model member out-of-straightness when

$$P_u/P_{eL} < 0.15 \qquad\qquad\qquad (2.1)$$

where

$$P_{eL} = \pi^2 EI/L^2. \tag{2.2}$$

For any axially loaded member where $P_u/P_{eL} > 0.15$, model member out-of-straightness by a sinusoidal sweep with a maximum value of $L/1000$ at the member mid-section. The sweep should be modeled in the direction that increases the maximum moment in the member.

2.2.3 Beam-to-column connections

General Requirements: Connections may be idealized as pinned or rigid if this idealization provides an acceptably accurate characterization of the connection response and does not significantly affect the calculation of overall frame strength (ultimate load) and frame displacements. Panel zones should be explicitly included in the connection model when they significantly affect the forces or displacements of the frame.

Recommended Approach:
The nonlinear connection response should be modeled explicitly; moment-rotation relationships may be determined either
- analytically using any of the well documented models available in the literature, or
- directly using experimentally obtained moment rotation characteristics

2.3 Limit States

If it can be demonstrated that a member limit state is directly captured within the analysis, it is not necessary to perform a check of that limit state using the corresponding strength resistance equation. In a planar analysis, these include: column in-plane flexural buckling, beam yielding (plastic moment), and beam-column in-plane limit states involving general yielding and stability effects.

Limit states not directly captured within in the analysis must be explicitly checked.

2.4 Resistance Factors

General Requirements: If advanced analysis methods are to be used in the context of AISC (2005), resistance factors, ϕ, must be incorporated in the analysis/design process.

Recommended approach: Apply a factor of 0.9 to the yield strength, F_y, as well as to the elastic modulus, E.

2.5 Serviceability

2.5.1 Appropriate load levels

Based on research by Galambos and Ellingwood (1986), ASCE/SEI 7-05 (ASCE /SEI 2005) includes a serviceability load combination for gravity load plus wind:

$$D + 0.5L + 0.7W$$

The live loads may be reduced as appropriate.

ASCE/SEI 7-05 also provides recommendations for serviceability load combinations for use in checking gravity load conditions. These include

$$D + L$$

and

$$D + 0.5 \ (S, \ R \ or \ L_r)$$

for cases involving visually objectionable deformations, repairable cracking or other damage to interior finishes and other short-term effects.

2.5.2 Limits on inelasticity and plastic rotation

As with any serviceability criteria, it is left to the engineer's discretion to determine how much yielding may be tolerated under serviceability load combinations. Using the load combinations listed above, it would be reasonable to expect that a typical advanced analysis formulation should detect no member cross-section yielding when the structure is subjected to these serviceability load combinations.

2.6 Live Load Reduction

General Requirements: The ASCE/SEI 7-05 (ASCE/SEI 2005) live load reduction provisions are defined on a member-by-member basis, and thus are not easily applied in an advanced analysis. Different reduced live loads are specified in general for different adjacent members, e.g., the live load reduction applied to a beam at a given level is generally different than that of a column supporting this beam. All the loads are applied to the same system model in an advanced analysis. The ASCE/SEI 7-05 live load reduction needs to be applied in such a way that it:

- Satisfies joint equilibrium;
- Assures that displacements calculated by the analysis are consistent with the resulting internal member force distribution; and
- Avoids the use of superposition, which is generally not valid in a nonlinear analysis.

Recommended approach: Determine "compensating forces" that account for the imbalance between the smaller live load reductions in the beams and larger reductions in the columns supporting these beams and the beams in all floors above them (Ziemian and McGuire, 1992). These compensating forces may be calculated using the following three-step procedure:

1. Determine reduced live loads for the beams that comprise the primary structural system. The reduction factors used in this step are based on the floor area supported by the beams. Using a simple gravity load takedown analysis, the total beam live load force $P_{i,j}^{bm.reduced}$ transferred to each column i at a particular level j is calculated.

2. Repeat Step 1 using the full (unreduced) beam live loads and reducing the resulting column live loads by reduction factors based on the floor areas being supported by the columns. Note that since these floor areas include all levels above them, the live load forces $P_{i,j}^{col.reduced}$ will in general be smaller than the forces calculated in Step 1.

3. Calculate the compensating forces as the difference between the above two sets of forces. This should be done by starting at the top level of the structure and working downward, making sure to include any compensating forces that have already been applied to the upper portions of each column. For a structure with n levels, the compensating force applied to column i at level j may be calculated as

$$F_{i,j} = P_{i,j}^{bm.reduced} - P_{i,j}^{col.reduced} - \sum_{k=j+1}^{n} F_{i,k} \tag{2.3}$$

All structural analyses that include live load are performed by applying a combination of the reduced beam live loads (calculated in Step 1) and the compensating forces (applied upward, opposite of gravity) determined in Step 3, as shown in Figure 2.1. The resulting forces in the beams and columns reflect the intent of the ASCE/SEI 7-05 (ASCE 2005) live load reduction provisions. In all cases where factored load combinations are investigated, both the beam live loads and the compensating forces should be multiplied by the appropriate live load factors. A detailed example of this approach is provided by Ziemian and McGuire (1992).

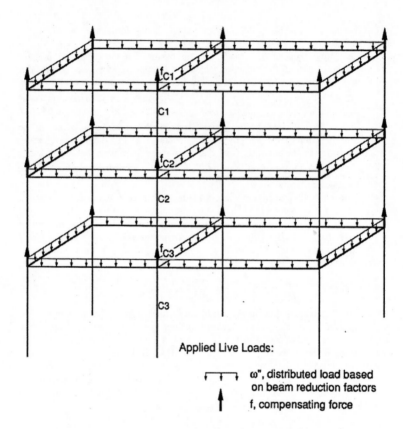

Figure 2.1 Application of compensating forces for live load reduction
Source: Ziemian and McGuire 1992. Copyright © American Institute of Steel
Construction; reproduced with permission from AISC.

PART 3:
BACKGROUND AND COMMENTARY

3.1 Background on Analysis Methods

Various methods of second-order inelastic analysis can potentially satisfy the requirements of Section 2.1.1. These approaches fall generally in one of two categories, as discussed below. Further discussion of the background to these models is provided in Section 3.2. The Methods are summarized in Table 3.1

3.1.1 Distributed plasticity methods

In a distributed plasticity analysis, the spread of yielding is tracked explicitly through the member cross-sections and along the member lengths. Stability effects are represented by expressing the equilibrium conditions on the inelastic deflected geometry. Inelastic stability effects and the spread of yielding are fundamentally coupled. This coupling and its influence on the maximum strength are particularly acute for cases such as weak-axis bending and axial compression on I-section members with intermediate unbraced lengths. Accurate consideration of these coupled effects by distributed plasticity analysis generally requires numerical computation using a large number of stress-strain sampling points through the cross-sections and along the member lengths. The variation of the inelastic displacements also must be sufficiently tracked along the member, e.g., by the use of a number of nodes along the length. Also, appropriate member nominal residual stresses, geometric imperfections and inelastic material stress-strain curves must be defined. Because of its fundamental tracking of the inelastic displacements, strains and stresses throughout the volume of the members, the distributed plasticity approach is the most rigorous and general method of second-order inelastic analysis.

For hot-rolled compact I-section members subjected to major-axis bending, ASCE (1997), Deierlein (2003), and Surovek-Maleck and White (2003 & 2004) have shown that distributed plasticity analysis closely replicates the in-plane AISC LRFD beam-column strengths based on an exact inelastic effective length, for a comprehensive range of end conditions and when the following nominal geometric imperfections, residual stresses and material idealizations are included in the analysis:

- A sinusoidal or parabolic out-of-straightness with a maximum amplitude of $\delta_o = L/1000$, where L is the unsupported length in the plane of bending.
- An out-of-plumbness of $\Delta_o = L/500$, the maximum tolerance specified in the AISC (2005) Code of Standard Practice.
- The Lehigh (Galambos and Ketter 1959) residual stress pattern shown in Fig. 3.1.
- An elastic-perfectly plastic material stress-strain response.

13

Table 3.1 Comparison of Analysis Methods

	Distributed Plasticity	Refined Plastic Hinge Model	
		Modified Tangent Modulus Approach	Direct Elastic-plastic Hinge Model
Modeling	Fiber discretization at cross-sections	Plastic hinges at element ends, otherwise elastic everywhere else	Plastic hinges at element ends, otherwise elastic everywhere else
Source of Inelasticity	Tracked explicitly at member-cross-sections and along the member	Tracked only at element ends	Tracked only at element ends
Spread of Inelasticity	Gradual development of inelasticity on cross-section and along member lengths	No spread of plasticity. An elastic–perfectly plastic hinge model to account for a full cross section yielding.	No spread of plasticity. An elastic–perfectly plastic hinge model to account for a full cross section yielding.
Material Nonlinear Behavior	Elastic – plastic model with or without strain hardening. Defined for each fiber.	The elastic-perfectly plastic hinge idealization enhanced by using equivalent reduced member flexural rigidities EI_x and EI_y	Beam-column strength interaction equation in AISC 2005 ($M_n = M_p$ and P_n calculated using actual member unsupported length).
Residual Stresses	Stress patterns defined uniquely at each fiber	Implicitly introduced into the plastic hinge idealization	Accounted for though t_b
Initial Out-of-Plumbness	Either introduced directly into analytical model or represented with notional loads	Either introduced directly into analytical model or represented with notional loads	Either introduced directly into analytical model or represented with notional loads
Flexural Rigidity Modification	None	Considered through use of the tangent modulus	Reduced by $0.9\tau_b$ (major-axis bending) or $0.8\tau_b$ (for minor axis bending) to account for distributed yielding effects
Computation Overhead	High	Lower	Lowest

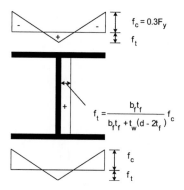

Fig. 3.1. Lehigh (Galambos and Ketter 1959) residual stress pattern.

These results are not surprising, since the AISC (2005) beam-column interaction equations were originally developed in part based on calibration to results from this type of analysis (ASCE 1997; Surovek-Maleck and White 2004). The above idealizations are sufficient to represent the in-plane resistances for other steel members having general compact doubly or singly-symmetric cross-sections, provided that an appropriate alternative nominal residual stress pattern is considered. Section 2.2.1 discusses appropriate residual stress patterns for different cross-section shapes. The 2005 AISC Specification uses the same nominal resistances and beam-column strength curves for rolled and welded I-section members. For members in which the governing limit states are yielding in bending and axial compression, including stability effects, the nominal flexural resistances are generally $M_n = M_p$, the same column resistance curve is employed for the nominal axial compression resistance P_n, and the same beam-column strength curve is employed. Therefore, advanced analyses conducted using the above nominal residual stress pattern are consistent with the Specification strength equations for either rolled or welded I-section members.

Distributed plasticity analysis, using the nominal attributes described above, forms the basis of the benchmark problems provided in Section 4, although various refinements are possible. For instance, different nominal residual stress distributions, as well as more comprehensive stress-strain models (e.g. models including strain hardening) may be incorporated in a distributed plasticity analysis. The above nominal attributes fully satisfy the base requirements of the AISC (2010) Specification for calculation of nominal in-plane column, beam and beam-column strengths for steel members.

3.1.2 Refined plastic hinge methods
Refined plastic hinge methods of analysis capture the detailed responses determined in the above distributed plasticity methods using overall member force-deformation,

moment-rotation or axial force-moment-curvature approximations. Many different approximations that have been proposed in the literature are acceptable. The only way to ascertain the acceptability of a given method is to subject it to the comprehensive benchmark problems discussed in Section 4. Two example of refined plastic hinge methods are discussed below: (1) the *Modified Tangent Modulus Approach* (Ziemian and McGuire 2002), and (2) the *Direct Elastic-Plastic Hinge Approach* (White et al. 2006).

3.1.2.1 Modified tangent modulus approach

In the Modified Tangent Modulus Approach, the base beam-column elastic-plastic hinge idealization is enhanced by using equivalent reduced member flexural rigidities EI_x and EI_y that vary as a function of the axial force and the minor-axis bending moment. The base beam-column elastic plastic hinge model is formulated using a fully-plastic strength envelope that represents the cross-section strength interaction between the axial force and biaxial bending for doubly-symmetric I-section members, assuming an elastic-perfectly plastic material. The stability effects are captured by a displacement-based formulation adopting a cubic variation of the transverse displacements between the nodes of the frame elements (McGuire et al. 2000). For cross-sections where the strength envelope is reached, the plastic interaction between axial force and the bending moments is represented using metal plasticity concepts. The member equivalent *major-axis* bending rigidity is defined as

$$EI_x^* = EI_x$$
$$\text{for} \quad P/P_y \leq 0.5 \tag{3.1a}$$

$$EI_x^* = EI_x\left(1+2\frac{P}{P_y}\right)\left(1-\frac{P}{P_y}\right)$$
$$\text{for} \quad P/P_y > 0.5 \tag{3.1b}$$

where EI_x^* is the equivalent bending rigidity, taken as uniform along the entire member length, EI_x is the cross-section elastic bending rigidity, and $P_y = A_g F_y$ is the cross-section yield load. For high axial load levels, this idealization is a close approximation of the AISC (2005) column inelastic stiffness reduction factor

$$\tau_a = 1$$
$$\text{for} \ P/P_y \leq 0.39 \tag{3.2a}$$

$$\tau_a = -2.743\frac{P}{P_y}\ln\left(\frac{P}{P_y}\right)$$
$$\text{for} \ P/P_y > 0.39 \tag{3.2b}$$

Indeed, for P/P_y slightly larger than 0.5 it falls between this stiffness reduction factor and the traditional CRC parabolic column stiffness reduction factor (Galambos 1998),

$\tau_b = 1$

for $P/P_y \leq 0.5$

(3.3a)

$$\tau_b = 4\frac{P}{P_y}\left(1-\frac{P}{P_y}\right)$$

for $P/P_y < 0.5$

(3.3b)

appearing in AISC (2005).

The member equivalent minor-axis bending rigidity of a given cross-section is defined as

$$EI_y^* =$$

$$\min\left\{EI_y, EI_y(1+2p)\left(1-\left[p+0.65\frac{M_{uy}}{M_{py}}\right]\right)\right\}$$

(3.4a)

where

$$p = \max\left(\frac{P}{P_y}, \ 0.25-0.325\frac{M_{uy}}{M_{py}}\right)$$

(3.4b)

In their matrix analysis, Ziemian and McGuire (2002) evaluate Eqs. (3.4) at the ends of the frame elements and assume a linear variation of the inelastic flexural rigidity between these points to obtain a closed-form stiffness matrix. This idealization accounts for the significant influence of minor-axis bending on the spread of yielding along the member length for doubly-symmetric steel I-section members.

The above approximations adequately capture the in-plane inelastic flexural stiffness of doubly-symmetric I-section members. Combined with the modeling of geometric imperfections discussed in Section 2.2.2, these procedures provide an accurate estimate of the resistances obtained from distributed plasticity analysis and satisfy the benchmarking requirements of Section 4.

3.1.2.2 Direct elastic-plastic hinge approach

Since 1961, the AISC Specifications have permitted the use of plastic analysis and design in cases where members containing plastic hinges satisfy requirements that ensure their ductility. However, the AISC Specifications from 1969 through 1999 have also required the engineer to supplement plastic analysis by beam-column strength interaction checks in which the axial resistance term is based on a member effective length. This practice adds significant complexity to the AISC plastic analysis and design procedures. Furthermore the resulting beam-column interaction equations provide, at best, only an approximate assessment of the frame stability behavior under progressive plastic hinge formation. In many cases, they overly restrict the forces and moments in sway frame columns (Ziemian et al. 1992).

The AISC (2010) Direct Analysis Method can be extended to provide an attractive alternative to the above procedures. The extension is very simple; for members that satisfy separate requirements to ensure sufficiently ductile response (i.e., sufficient rotation capacity), moment redistribution is allowed based on the assumption of elastic-perfectly plastic hinge behavior at the limit of member resistance given by the AISC (2005) beam-column interaction equations. This extension satisfies the AISC (2005) Appendix 1 provisions for inelastic analysis and design. The separate ductility requirements in AISC (2005) Appendix 1 include restrictions on:

- The material yield strength F_y,
- The flange and web slenderness values $b_f/2t_f$ and h_p/t_w,
- The member out-of-plane lateral-torsional buckling slenderness L_b/r_y,
- The magnitude of the axial force P_u and
- The connection details.

These restrictions limit the characteristics of beams and beam-columns such that their strengths are predicted accurately neglecting local and lateral-torsional buckling limit states. Therefore, for members that meet the above requirements, the "plastic hinging" strength is accurately approximated by the AISC (2005) beam-column interaction equations with $M_n = M_p$ and P_n calculated using the actual member unsupported length in the plane of bending. Furthermore, when P_u is less than $0.1P_{eL}$, where $P_{eL} = \pi^2 EI/L^2$ in the plane of bending, $P_n = P_y$ is an acceptable approximation (White et al. 2006). Members that do not satisfy all the requirements necessary to ensure ductile response may be designed elastically using the base Direct Analysis provisions of the AISC Specification .

The above type of analysis and design is referred to as the *Direct Elastic-Plastic Hinge Method*. In this method, the elastic flexural rigidity is reduced by $0.9\tau_b$ to account for distributed yielding effects neglected in the elastic-plastic hinge idealization, where τ_b is the traditional CRC parabolic column inelastic stiffness reduction factor, given by Eqs. (3.3). Also, a nominal initial out-of-plumbness (based on Section 2.2.2.1) is included to account for geometric imperfection effects. These devices eliminate the need to calculate and apply column effective lengths in the context of inelastic design, as long as the second-order effects are captured in the elastic-plastic hinge analysis. Furthermore, these devices allow the engineer to take advantage of basic second-order elastic-plastic hinge analysis software that is becoming increasingly more available in engineering practice.

For strong-axis bending of I-shaped members, benchmark studies have indicated maximum unconservative errors of 6 % using the AISC (2010) (elastic) DM versus 8 % using the traditional Effective Length Method (ELM) (Surovek-Maleck and White 2003) relative to distributed plasticity results. Note that the results using the Direct Elastic-Plastic Hinge Method are the same as those by the (elastic) Direct Analysis Method for individual member and non-redundant frame benchmarks.

To satisfy the Section 2.1.1 requirements for cases involving weak-axis bending and axial compression, the member elastic flexural rigidity must be reduced by 0.8τ in the direct elastic-plastic hinge method, where τ is the AISC column inelastic stiffness reduction factor given by Eqs. (3.2) (Surovek-Maleck and White 2004). The AISC (2010) (elastic) Direct Analysis Method allows larger than 6 % unconservative error relative to the worst-case distributed plasticity benchmark solutions for weak-axis bending. If $0.9\tau_b$ is used for the analysis, the worst-case unconservative error in the Direct Analysis and Direct Elastic-Plastic Hinge Methods is 13 %. However, the unconservative error is smaller than 6 % in most cases. As indicated above, for strong-axis bending, the errors are the same for the Direct Elastic-Plastic Hinge Analysis Method in basic individual member and non-redundant frame benchmarks.

It should be emphasized that, when using any of the above advanced inelastic analysis methods, inelastic redistribution shall not be allowed from any members that do not satisfy the Appendix 1 ductility requirements. Members that do not satisfy these requirements must be designed elastically to satisfy the AISC (2005) strength requirements based on the corresponding forces obtained from the inelastic analysis.

3.2 Background on Inelasticity

Incorporation of inelastic material behavior is necessary to obtain a more realistic approximation of the structural response of a frame. Depending on the level of accuracy required, as well as on the time and computational effort affordable, inelastic effects can be modeled in a reasonably accurate fashion. In the context of the inelastic analysis and design of steel frames, the plasticity models employed can be divided into two main groups, namely (i) distributed plasticity, also referred to as or spread-of-plasticity or plastic zone models and (ii) concentrated plasticity or plastic-hinge models. The features, advantages and limitations of these models are briefly described in the sections below.

3.2.1 Inelasticity in distributed plasticity models

Distributed plasticity models are considered to be the most accurate of analysis methods that utilize member elements to the fact that they make it possible to monitor and take into account) the spread of yielding both along the member length and throughout its cross-section. In particular, this feature enables the automatic handling of the interaction between stability and plasticity effects in a second-order analysis. However, the spread of plasticity can only be adequately modeled provided that the inelastic strain is tracked closely enough; this requires the consideration of a sufficiently large number of integration points within each member or finite element volume, which increases the computational effort when compared to a concentrated plasticity model. Thus, it is fair to say that distributed plasticity models are currently very rarely adopted in routine applications – instead, they are employed to prepare benchmark problems, subsequently used to validate and calibrate simpler and easier-to-use approaches such as those based on plastic-hinge models.

In the context of the analysis of frames subjected to in-plane bending, the members are usually viewed as assemblies of longitudinal "fibers". This automatically implies a uniaxial state of stress. Although this approach constitutes a simplification, since it disregards the influence of shear stresses on yielding, it has been found to provide sufficiently accurate results when compared to theoretical solutions or finite element results.

A distributed plasticity model requires both a uniaxial steel stress strain curve and a section residual stress pattern to implement. The most commonly used constitutive model is an elastic-perfectly plastic material model that neglects the effects of strain hardening. The yield plateau is often modeled with a nominal stiffness to prevent numerical difficulties resulting from a zero stiffness. Nevertheless, it is possible to consider more complete models that include strain-hardening effects. These models make it possible to take advantage of the additional steel strength. However, complex considerations involving local and overall member inelastic buckling are necessary to determine the maximum limit of the member resistance including strain-hardening effects.

Residual stresses are generally assumed to be longitudinally uniform, which means that only the self-equilibrated cross-section pattern is necessary. This residual stress pattern can either be based on experimental measurements (the most rigorous approach) or can consist of an idealized model, practically always involving linear or parabolic wall stress distributions.

3.2.2   Inelasticity in concentrated plasticity (plastic hinge) models

The concentrated plasticity model assumes the member inelastic behavior occurs only in a discrete number of cross-sections and also that these cross-sections yield instantaneously whenever their plastic strengths are reached. In the context of the in-plane analysis of planar frames, this plastic strength is a function of the axial force and bending moment, and the effect of the shear force is practically always neglected. Strain-hardening is not considered, and, the behavior of a cross-section is either elastic or perfectly plastic.) This has obvious numerical advantages; the frame exhibits an elastic behavior between the formation of any plastic hinges, which means that no numerical integrations are required to determine the frame behavior. The basic differences between the distributed and concentrated plasticity models lie in the fact that the latter (i) views the members as assembly of small segments each one represented by its cross-section, and (ii) does not capture the spread of plasticity either along the member length or through the member cross-sections.

The elastic-plastic analysis consists of a sequence of elastic analyses, which are carried out on "different frames", in the sense that they exhibit a growing number of plastic hinges. This means that (i) the frame inelastic stiffness and strength is generally overestimated since yielded fibers are assumed to continue behaving elastically, (ii) the interaction between stability and plasticity effects along the member's length is not properly modeled since the members are assumed to behave elastically between plastic hinge locations and (iii) the residual stress effects are not taken into account.

In order to overcome the above limitations, while retaining the numerical advantages of the plastic hinge approach, a number of refined plastic hinge methods of analysis have been proposed in the literature. Basically, these methods indirectly incorporate the spread of plasticity and residual stress effects into the plastic-hinge model by reducing the elastic stiffness properties of the frame members. Moreover, their validation and calibration is made by resorting to the results from benchmark solutions obtained through plastic zone analysis. Detailed explanations of two of these methods were presented in Section 3.1.

The presence of residual stresses in steel members is due to the non-uniform heating and cooling processes during fabrication (e.g., in hot-rolled members the larger mass of material located in the vicinity of the web-flange junctions cools slower than the remaining parts of the cross-section which leads to the presence of tensile normal residual stresses in these regions). Since the magnitude and distribution of the residual stresses locked in a given steel member depend on its cross-section geometry and manufacturing process, it is not possible to assign one common, idealized residual stress pattern=to all of them. Instead, there are a number of established residual stress patterns, each one associated to a given domain of application (*e.g.*, wide flange or welded I-section members). Therefore, strictly speaking, the residual stress distribution that best corresponds to the particular member cross-section geometry and fabrication procedures should be included in the model.

In the US, the residual stress pattern commonly used for hot-rolled wide flange I-sections is the one that was developed at Lehigh by Galambos and Ketter (1959), which consists of a bilinear stress distribution in the flanges combined with constant tensile stresses in the web. The maximum compressive stresses, occurring at the flange tips, are taken equal to *30%* of the yield stress (see Fig. 3.1). However, when the depth of the web is large, the assumption of constant residual stresses in the web becomes unrealistic, since the web central region cools considerably ahead of the web-flange junctions. This led to the use of alternative residual stress patterns in Europe, consisting of bilinear stress distributions both in the flanges and in the web, with equal maximum compressive and tensile stresses. The maximum values are specified equal to either *30%* or *50%* of the yield stress, depending on whether the web-to-flange width ratio is above or below 1.2 – i.e., for wide and narrow flange I-sections.

3.3 Background on Imperfections

It is generally acknowledged that the geometric imperfection effects included in a second-order inelastic analysis should represent the physical imperfections that may occur in the erected structure. The AISC Code of Standard Practice (AISC 2005) specifies the following tolerances:
- member out of straightness of $L/1000$

- for buildings with less than 20 stories, out-of-plumbness of H/500 in any shipping piece with a maximum lean, over the building height, of 1" towards the exterior or 2" towards the interior.

Imperfections associated with member out-of-plumbness are due to erection tolerances and include both individual column out-of plumbness and overall frame nonverticality. For low to medium rise buildings, it is appropriate to model the tolerances specified in the Code of Standard Practice explicitly as a uniform H/500 out-of-plumbness in a single direction (see Fig 3.2) – the one most detrimental to the overall stability of the structure.

A number of approaches have been suggested in the literature for modeling nonverticality. ECCS (1984) prescribes a reduction in out-of-plumbness based on the number of columns in a story and the number of stories in the building. However, Bridge & Bizzanelli (1997) considered the actual imperfections present in a 47 story office building, and their statistical data appears to contradict the ECCS provisions in that the imperfection values showed no correlation to the number of columns in a story.

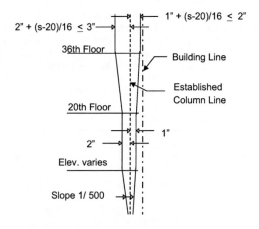

Figure 3.2 AISC specified allowable erection tolerances for building frames

In taller buildings, above approximately seven stories, a global nonverticality of H/500 exceeds the specified erection tolerance and can cause overly conservative results, by overestimating the P-Δ effects in the system. The two methods recommended for dealing with taller structures are based on studies by Bridge (1998) and Maleck and White (1998). Bridge suggests a uniform nonverticality of e_{oh}/H over the full height of the structure, where e_{oh} is the maximum permitted nonverticality. This approach suggests that the overall effect of imperfections in tall buildings can be captured as a cumulative effect over the height of the building, rather than being dominated by a "critical story" – this finding was later supported by Maleck (2001). However, Maleck and White (1998) suggested a more conservative approach: to model a "worst case" imperfection of H/500 over a portion of the building height, rather than to model a lower imperfection level over the entire height of the building.

The 6th edition of the SSRC stability guide (Ziemian 2010) states that member out-of-straightness only needs to be directly modeled in an advanced analysis in the event that it has a significant effect on frame behavior. It further suggests that the effect of out-of-straightness on frame behavior is based on (1) the relative magnitude of the member applied axial force and primary bending levels (2) whether the primary moments cause single or reverse curvature bending, and (3) the slenderness of the member. White & Nukala (1997) suggest that a limit of

$$P_u/P_{eL} < 1/7 \tag{3.5}$$

where

$$P_{eL} = \pi^2 EI/L^2 \tag{3.6}$$

is sufficient to restrict the reduction in strength due to out-of-straightness to less than 5% for a wide range of section types. In unbraced moment frames, the beam-columns are rarely loaded beyond this limit and direct modeling of member out-of-straightness can typically be neglected. In members that do require explicit modeling of the out-of-straightness, it is appropriate to use a sinusoidal or parabolic shape with a maximum value of L/1000 at the center, as specified by the AISC Code of Standard Practice (AISC 2005).

It is possible to incorporate the modeling of initial out-of-straightness directly within beam-column elements in advanced analysis software, i.e., using a curved element formulation. The software can automatically impose these imperfections in the direction in which the members are bent due to the loading applied on the structure.

3.4 Background on Connection Modeling

The ability to model connections as either rigid or pinned in an analysis is advantageous to the structural engineer. However, the use of advanced analysis implicitly requires that the effects of connection stiffness on the distribution of member forces (bending moments and axial forces) in a frame be modeled accurately. The structural engineer must therefore be confident that the connections assumed to be rigid in an advanced analysis model are sufficiently stiff to transmit "design" bending moments between connected members without appreciable moment shedding or underestimation of the second-order effects.

Due to the connection nonlinear response and the yielding of the frame members under the factored design load, a connection may be treated differently in the context of serviceability and strength limit state design procedures. For serviceability limit state design, the controlling parameter is the initial connection stiffness k_{ci}. For strength limit state design, the moment capacity of the connection M_{uc} must also be considered. Naturally, if a connection is treated as rigid in strength limit state design, it must equally qualify as a rigid connection for serviceability limit state design.

In order to be treated as rigid in an advanced analysis, braced frame connections typically require significantly lower stiffness values than those in comparable unbraced frames. Table 3.2 (adapted from Goto and Miyashita 1998) specifies the minimum initial rotational stiffness k_{ci} and moment capacity M_{uc} for a connection to be treated as rigid in an advanced analysis; only the minimum k_{ci} is relevant for serviceability limit state design. E is the elastic modulus, I_b is the second moment of area of the beam, L_b is the beam length, I_c is the second moment of area of the column, L_c is the column length, and M_{bp} is the plastic moment capacity of the beam.

Table 3.2 Minimum values for rigid connection parameters

	Unbraced Frame	Braced Frame
k_{ci}	$\dfrac{60EI_b}{L_b\left(1 + {I_b L_c}\big/{I_c L_b}\right)}$	$\dfrac{30EI_b}{L_b}$
M_{uc}	$\left\{(1.2+1.2\lambda)-(0.02+0.015\lambda)\dfrac{I_b L_c}{I_c L_b}\right\}M_{bp}$	$\left\{(0.75+0.9\lambda)-(0.01+0.015\lambda)\dfrac{I_b L_c}{I_c L_b}\right\}M_{bp}$

The parameter λ is the inelastic column slenderness ratio, defined as

$$\lambda = \frac{L_c}{r_c}\sqrt{\frac{F_y}{E}} \tag{3.6}$$

where r_c is the column radius of gyration in the plane of bending and F_y is the column yield stress.

Whenever possible, the full nonlinear response of a connection should be modeled directly in the analysis model. Some sources of data include connection databases, such as those developed by Goverdhan (1983) and Kishi and Chen (1986), and published results from the SAC-FEMA project (e.g. Roeder et al 2000, Liu and Astaneh-Asl 2000, Schneider and Teeraparbwong 2000, Swanson and Leon 2001). Phenomenological models have also proven to model connection behavior effectively (e.g. Kishi and Chen 1990). Any model that can be shown to effectively capture the nonlinear response of the connection may be considered.

3.5 Background on AISC Limit States

As noted in the introduction, any limit states not addressed comprehensively in the analysis must be accounted for through checks of member forces against resistance equations. The primary goal of a planar advanced analysis is to provide a direct assessment of in-plane system and compact-section member strengths. This is achieved by accounting for all the factors that influence the in-plane limit states behavior of compact-section members within the analysis.

The advanced analysis procedures outlined in Section 2.1.2 capture the in-plane resistances of compact I-section members completely. Therefore, the in-plane beam, column and beam-column resistance checks are automatically satisfied if the analysis shows that the structure is able to withstand the design loading. The engineer does not need to perform any separate evaluation of the in-plane member resistances, as the limit states listed in Section 2.3.1 are captured within the analysis and do not need to be checked via resistance equations.

For a planar advanced analysis to be valid, all members that reach their maximum in-plane resistance and subsequently redistribute forces to other portions of the structure must satisfy the AISC (2005) Appendix 1 requirements for inelastic analysis and design. These requirements are intended to ensure adequate ductility of the members such that the inelastic stiffness values and internal forces predicted by the planar advanced analysis models are sufficiently accurate. Moreover, it is necessary to check the out-of-plane beam-column resistances using the provisions of AISC (2005) Chapter H.

Members that do not satisfy the Appendix 1 requirements may be checked using the applicable column, beam, or beam-column resistance equations of the AISC (2005) Chapters E, F, G, H and I.

3.6 Background on Resistance Factors

If the above advanced analysis methods are to be used in an AISC (2005) LRFD context, the resistance factors ϕ_c and ϕ_b must be incorporated in the analysis-design process. Both of these factors are equal to 0.9 for all steel member strength limit states in AISC (2005). If one considers a single isolated member, one way of incorporating these factors is to generate the nominal beam-column strength curves from the analysis, and then multiply both the abscissa and the ordinate of these curves by $\phi_c = \phi_b = 0.9$ to obtain the final member design resistances. However, identical results are obtained if *both* the yield strength F_y and the elastic modulus E are factored by 0.9. Conversely, if only the yield strength F_y is factored by 0.9, the design strengths are overestimated for very slender columns that fail by elastic buckling.

The factoring of both E and F_y by 0.9 up front is preferred, since this approach facilitates the general analysis and design of structural systems. Therefore, all the elastic stiffness values of the advanced analysis model must be multiplied by 0.9. Furthermore, all the properties of the analysis model that depend on the yield strength F_y (e.g., P_y, M_p, etc.) must also be multiplied by 0.9. Since there is no straightforward way of applying any form of inelastic analysis in the context of ASD, AISC (2010) disallows its use.

3.7 Background on Serviceability

Historically, the AISC serviceability provisions have been brief and have defined only general (non-specific) performance requirements. For example, the 2005 AISC Specification indicates that drift should be evaluated under service loads to provide for serviceability of the structure. References such as ASCE (1988), Griffis (1993) and Griffis and White (2007) provide guidance on appropriate inter-story drift limits. These limit states should also be checked when designing using advanced analysis.

Engineers use a wide range of loadings for assessing serviceability. In the case of wind loading, for example, engineers have often used nominal (code-specified) loads, e.g., from ASCE/SEI 7-05 (ASCE 2005). Others have used smaller values, as low as those corresponding to a 10 year recurrence interval for certain types of structures (Griffis, 1993).

In addition to the above checks for serviceability, when using advanced analysis it is important to perform further checks to ensure adequate performance of the structure under serviceability load combinations. Specifically, it is important to ensure minimal yielding of all structural components within the steel structures considering (1) the inelastic rotation in connections in general and in specific connection components and (2) the inelasticity in all girders, columns and bracing members.

3.8 Background on Live Load Reduction

One of the primary benefits of using advanced methods of inelastic analysis within the design process is the fact that it grants the opportunity to capture inelastic force redistribution. However, this opportunity commits the Engineer to modeling the entire structural system or, at least, a substantial portion of it. In addition to correctly modeling the stiffness and strength of the system, an essential part of the analysis is an accurate representation of the applied loadings.

In most design load standards, the effects of live load are often reduced to reflect the low probability of all live load occurring simultaneously at all or a substantial portion of the structure. Obviously, such reductions can have a significant influence on a structure's response. For example, reduced live loads will most likely produce smaller second-order effects. It is important to note that this does not imply that not reducing live loads is always conservative. To understand this fact, one may consider the case of a structure resisting an overturning moment produced by lateral load. In this case, using a reduced live load can be unconservative.

Live load reduction provisions provided in ASCE7-05 (ASCE 2005) are defined on a member-by-member basis. For members such as columns that support a large floor area, there is often a significant reduction in the members' live load. In contrast, the reduction for beams supporting a comparatively smaller amount of area may be little to none. In this regard, it is apparent that these provisions were developed for the sole intent of member proportioning. In other words, they are ideally suited for a member analysis and design approach where the live load effect is calculated separately, reduced accordingly, and then added to other load effects by applying the principle of superposition. This can only be done with the results of a first-order elastic analysis. Another drawback of this approach is that the member design forces are often not consistent with the resulting deflections calculated by a structural analysis. One such method to alleviate these issues (Ziemian and McGuire 1992) has been included as a recommended approach in these design rules.

PART 4:
BENCHMARKS

While many commercial software packages currently provide some method for performing second-order analysis, few (if any) have the robust inelastic analysis capabilities to be used in an advanced analysis application. In order to allow the engineer to validate emerging computer algorithms or software packages for accuracy, as well as to determine their limitations, representative benchmark solutions are provided herein. Each benchmark includes both graphical and tabulated data.

Numerous references are available that provide benchmarks of second-order inelastic analysis of columns and two-dimensional frames. Those provided by Kanchanalai (1977) and Vogel (1985) are of primary interest having been used as the basis for specification strength equations. The Kanchanalai solutions are based on perfect frame geometry and were used by AISC when developing the beam-column interaction equations first published in the 1986 AISC LRFD Specification. ECCS adopted the Vogel frames as benchmark solutions for second-order inelastic analysis. Besides the studies included in this report, almost all of which have been previously reported, the reader is referred to White and Chen (1993), Ziemian (1993), Clarke et al (1993), Lui (1993), Maleck (2001), Martinez-Garcia (2002), and the Commentary of the AISC Specification (AISC 2010), among others, for additional benchmark solutions.

The columns and frame benchmarks below were analyzed using distributed plasticity analyses (see Section 3.1.1) with the Galambos and Ketter (1959) residual stress pattern explicitly modeled. Initial out-of-plumbness was explicitly modeled with a value of L/500, and initial out-of-straightness was explicitly modeled in beam-columns contributing to the lateral resisting system with a sinusoidal sweep of L/1000.

4.1 Benchmark Columns

The benchmark solutions are limited to mid-range rolled I-shapes. The columns are loaded non-proportionally; a predetermined axial load was applied first followed by application of a lateral load to the point of failure. Two sets of data are provided: normalized axial load (P/P_y) versus the first-order moment due to the applied lateral load (HL/M_p or $2HL/M_p$) and the normalized axial load versus the calculated second-order moment due to the applied load. The benchmark solutions provided were used in verification of the Direct Analysis Method for the 2005 AISC Specification (Surovek-Maleck and White 2003) The solutions presented are nominal, i.e., neither the yield strength nor the elastic stiffness has been factored.

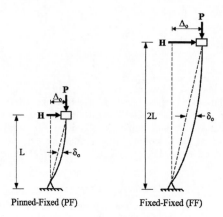

Figure 4.1 Column Benchmark Cases
Source: Maleck 2001.

The columns for which solutions are provided are shown in Figure 4.1. Column designations are provided as follows: Fixity_slenderness_strong-axs or weak-axis bending, (e.g. PF40S is a pinned-fixed column, L/r = 40 in strong-axis bending.) The slenderness ratio given is about the axis of bending.

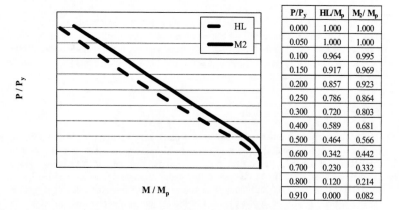

P/P_y	HL/M_p	M_2/M_p
0.000	1.000	1.000
0.050	1.000	1.000
0.100	0.964	0.995
0.150	0.917	0.969
0.200	0.857	0.923
0.250	0.786	0.864
0.300	0.720	0.803
0.400	0.589	0.681
0.500	0.464	0.566
0.600	0.342	0.442
0.700	0.230	0.332
0.800	0.120	0.214
0.910	0.000	0.082

Figure 4.2 Benchmark Column PF_20S

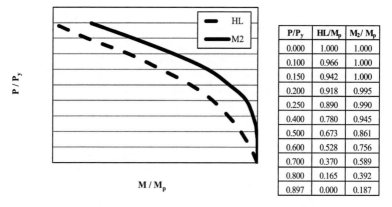

P/P_y	HL/M_p	M_2/M_p
0.000	1.000	1.000
0.100	0.966	1.000
0.150	0.942	1.000
0.200	0.918	0.995
0.250	0.890	0.990
0.400	0.780	0.945
0.500	0.673	0.861
0.600	0.528	0.756
0.700	0.370	0.589
0.800	0.165	0.392
0.897	0.000	0.187

Figure 4.3 Benchmark Column PF_20W

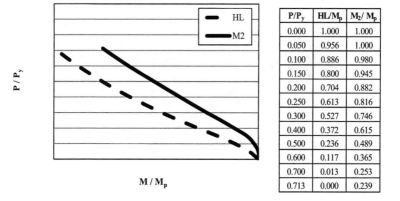

P/P_y	HL/M_p	M_2/M_p
0.000	1.000	1.000
0.050	0.956	1.000
0.100	0.886	0.980
0.150	0.800	0.945
0.200	0.704	0.882
0.250	0.613	0.816
0.300	0.527	0.746
0.400	0.372	0.615
0.500	0.236	0.489
0.600	0.117	0.365
0.700	0.013	0.253
0.713	0.000	0.239

Figure 4.4 Benchmark Column PF_40S

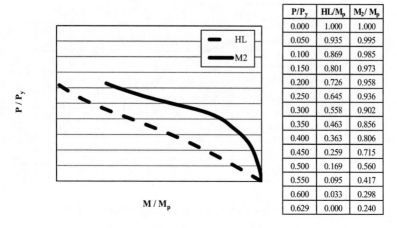

P/P$_y$	HL/M$_p$	M$_2$/ M$_p$
0.000	1.000	1.000
0.050	0.935	0.995
0.100	0.869	0.985
0.150	0.801	0.973
0.200	0.726	0.958
0.250	0.645	0.936
0.300	0.558	0.902
0.350	0.463	0.856
0.400	0.363	0.806
0.450	0.259	0.715
0.500	0.169	0.560
0.550	0.095	0.417
0.600	0.033	0.298
0.629	0.000	0.240

Figure 4.5 Benchmark Column PF_40W

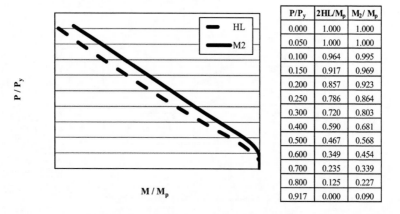

P/P$_y$	2HL/M$_p$	M$_2$/ M$_p$
0.000	1.000	1.000
0.050	1.000	1.000
0.100	0.964	0.995
0.150	0.917	0.969
0.200	0.857	0.923
0.250	0.786	0.864
0.300	0.720	0.803
0.400	0.590	0.681
0.500	0.467	0.568
0.600	0.349	0.454
0.700	0.235	0.339
0.800	0.125	0.227
0.917	0.000	0.090

Figure 4.6 Benchmark Column FF_20S

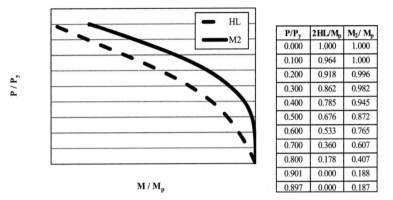

P/P$_y$	2HL/M$_p$	M$_2$/ M$_p$
0.000	1.000	1.000
0.100	0.964	1.000
0.200	0.918	0.996
0.300	0.862	0.982
0.400	0.785	0.945
0.500	0.676	0.872
0.600	0.533	0.765
0.700	0.360	0.607
0.800	0.178	0.407
0.901	0.000	0.188
0.897	0.000	0.187

Figure 4.7 Benchmark Column FF_20W

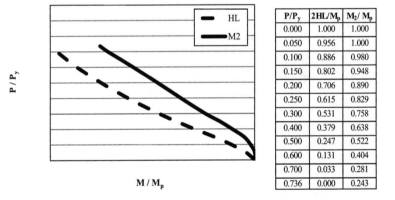

P/P$_y$	2HL/M$_p$	M$_2$/ M$_p$
0.000	1.000	1.000
0.050	0.956	1.000
0.100	0.886	0.980
0.150	0.802	0.948
0.200	0.706	0.890
0.250	0.615	0.829
0.300	0.531	0.758
0.400	0.379	0.638
0.500	0.247	0.522
0.600	0.131	0.404
0.700	0.033	0.281
0.736	0.000	0.243

Figure 4.8 Benchmark Column FF_40S

34   ADVANCED ANALYSIS IN STEEL FRAME DESIGN

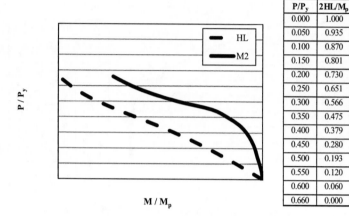

P/P$_y$	2HL/M$_p$	M$_2$/ M$_p$
0.000	1.000	1.000
0.050	0.935	0.995
0.100	0.870	0.985
0.150	0.801	0.973
0.200	0.730	0.962
0.250	0.651	0.945
0.300	0.566	0.921
0.350	0.475	0.885
0.400	0.379	0.827
0.450	0.280	0.749
0.500	0.193	0.584
0.550	0.120	0.459
0.600	0.060	0.361
0.660	0.000	0.269

Figure 4.9 Benchmark Column FF_40W

4.2   Benchmark Frames: Nominal

All of the frame solutions presented herein have been collected from various studies documented in the literature. They are provided to allow for a means of examining different cases of planar framing systems analyzed using the approaches identified in this report. The frames range from simple portal frames to multi-story, multi-bay frames; they include symmetric and unsymmetric geometry and loading, as well as unbraced and combined braced and unbraced framing. In all instances, a reference to the original study is provided. Results for each frame case include:
- Load-deflection curves
- Tabulated data points for the load-deflection curves

4.2.1   Maleck Industrial Frame
The single-story frame shown in Figure 4.10 was initially developed by Maleck (2001) and studied further by Deierlein (2003), Kuchenbecker et al (2004), and White et al. (2006). It represents a frame typical of some single-story industrial buildings in which a few columns provide lateral support for a large number of bays (which may or may not be in the plane of the lateral frame). The drift limits on such buildings are usually quite liberal since the cladding is not sensitive to drift. Due to the gravity to wind load ratio and the liberal drift values, this frame behavior is largely dominated by P-Δ effects and is quite sensitive to the inclusion of an initial out-of-plumbness. The load deflection curve, shown in Figure 4.11, represents the response due to non-proportional loading in the gravity load-case (LC = 1.2D + 1.6L) . Tabulated P-Δ data are given in Table 4.1.

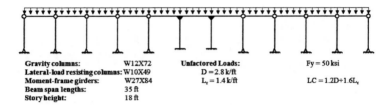

Gravity columns:	W12X72	Unfactored Loads:	Fy = 50 ksi
Lateral-load resisting columns:	W10X49	D = 2.8 k/ft	
Moment-frame girders:	W27X84	L_r = 1.4 k/ft	LC = 1.2D+1.6L_r
Beam span lengths:	35 ft		
Story height:	18 ft		

Figure 4.10 11-bay industrial building frame

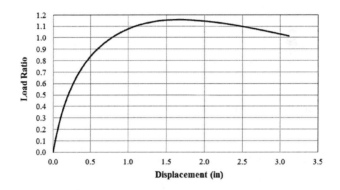

Figure 4.11 Load-deflection response, 11-bay industrial frame

Table 4.1 Load-deflection response, 11-bay industrial frame

Applied Load Ratio	Displacement (in)	Applied Load Ratio	Displacement (in)
0.000	0.000	1.07	0.970
0.128	0.038	1.09	1.07
0.253	0.082	1.12	1.19
0.372	0.133	1.14	1.32
0.484	0.191	1.15	1.44
0.587	0.257	1.16	1.59
0.681	0.331	1.16	1.75
0.764	0.411	1.15	1.93
0.836	0.497	1.14	2.12
0.898	0.587	1.12	2.35
0.952	0.680	1.08	2.63
0.998	0.775	1.02	3.11
1.04	0.872	1.02	3.11

4.2.2   Vogel Frame

One of the most commonly referenced frames is the one shown in Fig. 4.12, originally presented by Vogel (1985) as part of a series of benchmark frames for second-order inelastic analysis. Load-deflection curves are shown in Figure 4.13, and load-deflection data are provided in Table 4.2

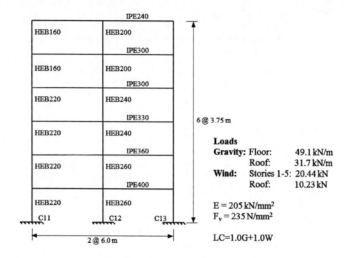

Figure 4.12 Vogel Frame

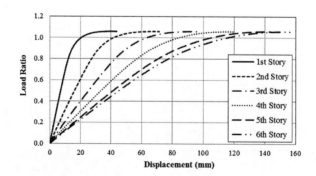

Figure 4.13 Load-deflection response, Vogel Frame

Table 4.2 Load-deflection data, Vogel Frame

Applied Load Ratio	Displacement (mm)					
	1st Story	2nd Story	3rd Story	4th Story	5th Story	6th Story
0.000	0.000	0.000	0.000	0.000	0.000	0.000
0.099	1.33	3.02	4.74	6.27	7.76	8.48
0.198	2.70	6.12	9.59	12.7	15.7	17.2
0.296	4.10	9.30	14.6	19.3	23.9	26.1
0.395	5.53	12.6	19.7	26.1	32.2	35.2
0.494	6.99	15.9	24.9	33.0	40.8	44.6
0.592	8.49	19.3	30.3	40.1	49.7	54.2
0.690	10.0	22.8	35.8	47.5	58.9	64.3
0.783	11.7	26.5	41.6	55.3	68.7	75.3
0.872	13.8	30.6	47.9	63.9	79.7	87.6
0.939	16.3	35.1	54.4	72.6	90.7	99.8
0.989	19.5	40.1	61.1	81.3	101	112
1.02	23.0	45.3	67.6	89.3	111	122
1.04	27.0	51.0	74.4	97.2	120	132
1.05	31.8	57.4	81.6	105	129	141
1.06	37.1	63.9	88.3	112	136	148
1.06	38.6	65.7	90.2	114	138	150
1.06	40.7	68.1	92.7	117	140	153
1.06	42.4	70.0	94.7	119	142	155
1.06	43.8	71.5	96.2	120	144	156

4.3   Benchmark Frames: Factored

The following two frames are included to provide results when both the material strength and stiffness are factored by 0.9, in accordance with the design guidelines presented in this report. The frames results were originally presented in Martinez-Garcia (2002).

4.3.1   Ziemian Frame

The two-story example shown in Figure 4.14 was originally studied by Ziemian (1990) as part of a large parametric study including both light and heavy gravity load levels. The load deflection curves are shown in Figure 4.15, and P-Δ data are provided in Table 4.3. It has an unusual load-deflection response; each story drifts in a different direction under gravity load. In addition, the lateral deflection shifts near the limit load as high levels of plastic deformation occur, changing the distribution of stiffness in the frame. It is often cited as a benchmark frame for considering effects of redistribution of forces near the limit load.

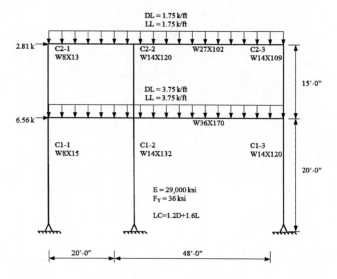

Figure 4.14 Ziemian frame

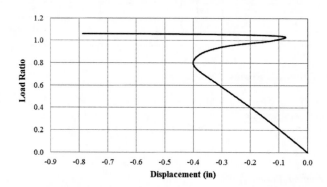

Figure 4.15 Second-story load-deflection response, Ziemian frame

Table 4.3 Load-deflection response, Ziemian frame

Applied Load Ratio	Displacement (in)	Applied Load Ratio	Displacement (in)	Applied Load Ratio	Displacement (in)
0.000	0.000	0.970	-0.226	1.05	-0.127
0.100	-0.046	0.971	-0.223	1.05	-0.153
0.199	-0.094	0.972	-0.220	1.05	-0.168
0.298	-0.144	0.972	-0.218	1.05	-0.187
0.397	-0.196	0.973	-0.217	1.05	-0.214
0.497	-0.250	0.973	-0.214	1.05	-0.246
0.596	-0.307	0.974	-0.210	1.06	-0.284
0.696	-0.363	0.976	-0.205	1.06	-0.330
0.740	-0.386	0.978	-0.197	1.06	-0.385
0.780	-0.399	0.981	-0.186	1.06	-0.449
0.810	-0.401	0.984	-0.171	1.06	-0.524
0.832	-0.397	0.989	-0.155	1.06	-0.612
0.851	-0.391	0.993	-0.144	1.06	-0.615
0.868	-0.383	0.997	-0.131	1.06	-0.620
0.885	-0.372	1.00	-0.119	1.06	-0.623
0.902	-0.358	1.01	-0.107	1.06	-0.627
0.917	-0.340	1.02	-0.094	1.06	-0.634
0.931	-0.320	1.02	-0.085	1.06	-0.643
0.942	-0.301	1.03	-0.078	1.06	-0.656
0.952	-0.279	1.03	-0.076	1.06	-0.674
0.961	-0.255	1.03	-0.078	1.06	-0.699
0.969	-0.230	1.04	-0.083	1.06	-0.736
0.969	-0.229	1.04	-0.093	1.06	-0.788
0.969	-0.228	1.04	-0.107		

4.3.2   Martinez-Garcia Moment Frame

The unbraced frame shown in Figure 4.16 was presented by Martinez-Garcia (2002). This frame highlights the difficulty in using design methods that are dependent on story-based stability assumptions, such as simultaneous story buckling. Load-deflection results for the 1.2D + 1.6L + 0.8W load case are presented in Figure 4.17. Loads were applied proportionally to the structure, and load deflection data are presented in Table 4.4.

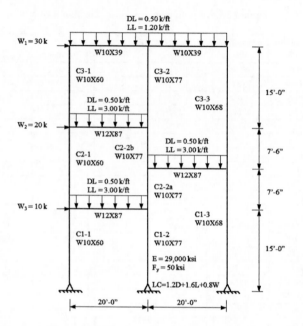

Figure 4.16 Martinez-Garcia Moment Frame

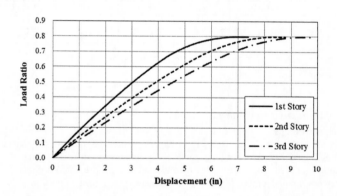

Figure 4.17 Load-deflection curves, Martinez-Garcia Moment Frame

Table 4.4 Load-deflection curves, Martinez-Garcia Moment Frame

Applied Load Ratio	Displacement (in)		
	1st Story	2nd Story	3rd Story
0.000	0.000	0.000	0.000
0.049	0.264	0.346	0.407
0.099	0.536	0.700	0.822
0.148	0.815	1.06	1.25
0.198	1.10	1.43	1.68
0.247	1.40	1.81	2.12
0.296	1.70	2.20	2.57
0.346	2.01	2.60	3.03
0.395	2.33	3.01	3.51
0.444	2.66	3.43	3.99
0.494	3.00	3.86	4.49
0.543	3.35	4.30	4.99
0.592	3.72	4.76	5.52
0.640	4.10	5.24	6.06
0.687	4.53	5.75	6.63
0.728	5.01	6.31	7.26
0.760	5.52	6.88	7.87
0.784	6.10	7.49	8.52
0.796	6.69	8.11	9.16
0.797	7.38	8.78	9.85

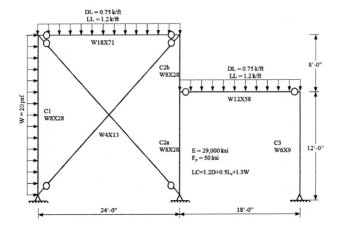

Figure 4.18 Martinez-Garcia Braced Frame (Strong Axis Bending)

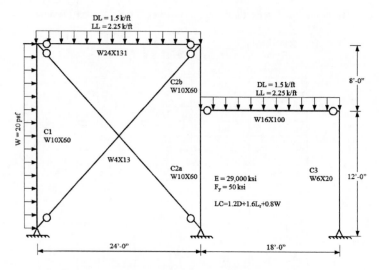

Figure 4.19 Martinez-Garcia Braced Frame (Weak Axis)

4.3.3 Martinez-Garcia Braced Frame

Figures 4.18 and 4.19 presents a braced frame presented by Martinez-Garcia (2002) in strong and weak axis bending; these frames also highlight difficulties with story-based stability design evaluation approaches. Load deflection response is provided in Figure 4.20 and 4.21 for the strong and weak-axes, respectively and tabulated data are presented in Tables 4.5 and 4.6.

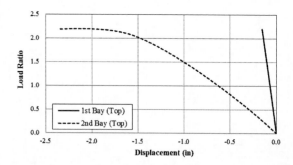

Figure 4.20 Load-deflection curves, Martinez-Garcia Braced Frame (Strong Axis)

Table 4.5 Load-deflection curves, Martinez-Garcia Braced Frame (Strong Axis Bending)

Applied Load Ratio	Displacement (in)		Applied Load Ratio	Displacement (in)	
	1st Bay (Top)	2nd Bay (Top)		1st Bay (Top)	2nd Bay (Top)
0.000	0.000	0.000	1.24	-0.085	-0.802
0.050	-0.003	-0.028	1.29	-0.088	-0.840
0.100	-0.007	-0.055	1.34	-0.092	-0.879
0.149	-0.010	-0.084	1.39	-0.095	-0.918
0.199	-0.013	-0.112	1.44	-0.099	-0.957
0.249	-0.017	-0.141	1.49	-0.102	-1.00
0.299	-0.020	-0.170	1.54	-0.106	-1.04
0.349	-0.023	-0.200	1.59	-0.109	-1.08
0.398	-0.027	-0.230	1.64	-0.113	-1.12
0.448	-0.030	-0.260	1.69	-0.116	-1.16
0.498	-0.033	-0.291	1.74	-0.120	-1.21
0.548	-0.037	-0.322	1.79	-0.123	-1.25
0.597	-0.040	-0.353	1.84	-0.127	-1.30
0.647	-0.044	-0.385	1.89	-0.130	-1.35
0.697	-0.047	-0.417	1.94	-0.134	-1.40
0.747	-0.050	-0.450	1.98	-0.137	-1.46
0.797	-0.054	-0.483	2.03	-0.141	-1.52
0.846	-0.057	-0.517	2.08	-0.144	-1.58
0.896	-0.061	-0.551	2.12	-0.148	-1.67
0.946	-0.064	-0.585	2.16	-0.150	-1.77
0.996	-0.068	-0.620	2.18	-0.152	-1.88
1.05	-0.071	-0.656	2.19	-0.154	-2.02
1.10	-0.074	-0.692	2.20	-0.154	-2.18
1.14	-0.078	-0.728	2.19	-0.154	-2.36
1.19	-0.081	-0.765			

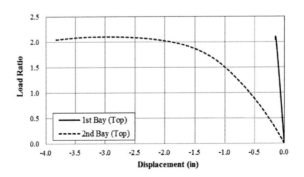

Figure 4.21 Load-deflection curves, Martinez-Garcia Braced Frame (Weak Axis)

Table 4.6 Load-deflection curves, Martinez-Garcia Braced Frame (Weak Axis)

Applied Load Ratio	Displacement (in)		Applied Load Ratio	Displacement (in)	
	1st Bay (Top)	2nd Bay (Top)		1st Bay (Top)	2nd Bay (Top)
0.000	0.000	0.000	1.19	-0.077	-0.730
0.050	-0.003	-0.024	1.24	-0.080	-0.770
0.099	-0.006	-0.048	1.29	-0.084	-0.811
0.149	-0.009	-0.073	1.34	-0.087	-0.853
0.199	-0.012	-0.098	1.39	-0.091	-0.896
0.248	-0.016	-0.124	1.44	-0.094	-0.940
0.298	-0.019	-0.150	1.49	-0.097	-0.986
0.347	-0.022	-0.177	1.53	-0.101	-1.03
0.397	-0.025	-0.204	1.58	-0.104	-1.08
0.446	-0.028	-0.232	1.63	-0.108	-1.14
0.496	-0.031	-0.260	1.68	-0.111	-1.19
0.546	-0.035	-0.289	1.73	-0.115	-1.26
0.595	-0.038	-0.319	1.77	-0.118	-1.33
0.645	-0.041	-0.349	1.82	-0.122	-1.41
0.694	-0.044	-0.380	1.87	-0.125	-1.50
0.744	-0.048	-0.411	1.91	-0.129	-1.60
0.793	-0.051	-0.444	1.96	-0.133	-1.73
0.843	-0.054	-0.477	2.00	-0.136	-1.89
0.892	-0.057	-0.510	2.04	-0.140	-2.10
0.942	-0.061	-0.545	2.08	-0.144	-2.37
0.991	-0.064	-0.580	2.10	-0.148	-2.68
1.04	-0.067	-0.616	2.10	-0.150	-3.02
1.09	-0.070	-0.653	2.09	-0.151	-3.40
1.14	-0.074	-0.691	2.05	-0.150	-3.83

REFERENCES

AISC (2010). *ANSI/AISC 360-10 Specification for Structural Steel Buildings*, American Institute of Steel Construction, Chicago, IL.

AISC (2005). *Steel Construction Manual, Thirteenth Edition*, American Institute of Steel Construction, Chicago, IL.

Alemdar, B. N. (2001). "Distributed plasticity analysis of steel building structural systems." PhD thesis, Georgia Institute of Technology, Atlanta.

Alemdar, B.N. and White, D.W. (2005), "Displacement, Flexibility, and Mixed Beam–Column Finite Element Formulations for Distributed Plasticity Analysis", Journal of Structural Engineering, Vol. 131, No. 12ASCE (2005). *Minimum Design Loads for Buildings and Other Structures*, ASCE/SEI 7-05, American Society of Civil Engineers, Reston, VA.

ASCE Task Committee on Drift Control of Steel Building Structures (1988). "Wind Drift Design of Steel-Framed Buildings: State of the Art", *Journal of the Structural Division*, ASCE, Vol. 114, No. 9, pp. 2085-2108.

ASCE (1997), *Effective Length and Notional Load Approaches for Assessing Frame Stability: Implications for American Steel Design*, American Society of Civil Engineers Structural Engineering Institute's Task Committee on Effective Length under the Technical Committee on Load and Resistance Factor Design, 442 pp.

ASCE/SEI (2005) (ASCE/SEI 7-05), *Minimum Design Loads for Buildings and Other Structures*, ASCE, Reston, VA.

Bridge, R.Q. and Bizzanelli, P. (1997). Imperfections in Steel Structures, *Proceedings - 1997 Annual Technical Session, and Meeting, Structural Stability Research Council*, pp. 447-458.

Bridge, R.Q. (1998). "The Inclusion of Imperfections in Probability-Based Limit States Design", *Proceedings of the 1998 Structural Engineering World Congress*, San Francisco, California, July.

Chen, W.F. and Kim, S.E. (1997). *LRFD Steel Design Using Advanced Analysis*, CRC Press, Boca Raton, FL, 467 pp.

Chen, W.F. and Toma, S., eds. (1994). Advanced Analysis of Steel Frames, CRC Press, 384 pp.

Clarke, M.J. and Bridge, R.Q. (1997). "*Notional Load Approach for the Assessment of Frame Stability*", Chapter 4, ASCE Committee Monograph, *Effective Length and Notional Load Approaches for Assessing Frame Stability: Implications for American Steel Design*, American Society of Civil Engineers Structural Engineering Institute's

Task Committee on Effective Length under the Technical Committee on Load and Resistance Factor Design, pp. 181-278.

Clarke, M.J., Bridge, R.Q., Hancock, G.J., and Trahair, N.S. (1992). "Advanced Analysis of Steel Building Frames", *Journal of Constructional Steel Research*, V. 23, no. 1-3, pp. 1–29.

Clarke, M.J., Bridge, R.Q., Hancock, G.J., and Trahair, N.S. (1993). "Benchmarking and Verification of Second-Order Elastic and Inelastic Frame Analysis Programs", in White, Donald W. and Chen, W. F., eds. (1993), *Plastic Hinge Based Methods for Advanced Analysis and design of Steel Frames – an assessment of the State of the Art*, Structural Stability research Council, Bethlehem, PA, 299 pp.

Deierlein, G., Hajjar, J.F., Yura, J.A., White, D.W., and Baker, W.F. (2002). "Proposed New Provisions for Frame Stability Using Second-Order Analysis", *Proceedings - 2002 Annual Technical Session, Structural Stability Research Council*, pp. 1-20.

Deierlein, G. (2003). "Background and Illustrative Examples on Proposed Direct Analysis Method for Stability Design of Moment Frames", *Background Materials*, AISC Committee on Specifications, Ballot 2003-4-360-2, August 20, 17 pp.

ECCS (1984). *Ultimate Limit States Calculations of Sway Frames With Rigid Joints*, Technical Working Group 8.2, Systems, Publications No. 33, European Convention For Constructional Steelwork, 20 pp.

Galambos, T.V. (ed.) (1998). *Guide to Stability Design Criteria for Metal Structures*, 5th Edition, Structural Stability Research Council, Wiley, 911 pp.

Galambos, T.V. and Ellingwood, B.R. (1986). "Serviceability Limit States: Deflections", *Journal of the Structural Division*, ASCE, Vol. 112, No. 1, pp. 67-84.

Galambos, T.V. and Ketter, R.L. (1959), "Columns Under Combined Bending and Thrust", *Journal of the Engineering Mechanics Division*, ASCE, 85(EM2), pp. 135-152.

Griffis, L.G. (1993). "Serviceability Limits States under Wind Load", *Engineering Journal*, AISC, Vol. 30, No. 1, pp. 1-16.

Griffis, L.G. and White, D.W. (2007). *Stability Design of Steel Buildings*, Steel Design Guide, American Institute of Steel Construction.

Goverdhan, A.V. (1983). *A Collection of Experimental Moment-Rotation Curves and Evaluation of Prediction Equations for Semi-Rigid Connections*, M.S. Thesis, Vanderbilt University, Nashville, TN.

Goto, Y., and Miyashita, S. (1998). "Classification system for rigid and semi-rigid connections", *Journal of Structural Engineering*, Vol. 17, No. 8, Sept. pp. 544-553.

Kaehler, R.C., White, D.W. and Kim, Y.D. (2007). *Frame Design Using Web-Tapered Members*, Steel Design Guide, Metal Building Manufacturers Association and American Institute of Steel Construction.

Kanchanalai, T. (1977). *The Design and Behavior of Beam-Columns in Unbraced Steel Frames*, AISI Project No. 189, Report No. 2, Civil Engineering/Structures Research Lab., University of Texas, Austin, TX, 300 pp.

Kishi, N., and Chen, W.F. (1986). *Database of Steel Beam-to-Column Connections*, Structural Engineering Report No. CE-STR-86-26, 2 vols., School of Civil Engineering, Purdue University, West Lafayette, IN.

Kishi, N., and Chen, W.F. (1990), "Moment-Rotation Relations for Semi-Rigid Connections with Angles," Journal of Structural Engineering, ASCE, Vol. 116, No. 7, pp. 1813-1834.

Kuchenbecker, G. H., White, D.W. and Surovek-Maleck, A.E. (2004), "Simplified Design of Building Frames using First-Order Analysis and K = 1.0," *Proceedings of the 2004 SSRC Annual Technical Sessions and Meeting*, Long Beach, March.

Lui, E. (1993). "Workshop Summary Report: Verification and Benchmark Problems", in White, Donald W. and Chen, W. F., eds. (1993), *Plastic Hinge Based Methods for Advanced Analysis and design of Steel Frames – an assessment of the State of the Art*, Structural Stability research Council, Bethlehem, PA, pp. 275 - 278.

Liu, J. and Astaneh-Asl, A. (2000), "*Cyclic Tests on Simple Connections, Including Effects of Slab*, Report No. SAC/BD-00/03, SAC Steel Project Background Document, SAC Joint Venture.

Maleck (Surovek), A.E., White, D.W. and Chen, W.F. (1995). "Practical Application of Advanced Analysis in Steel Design", Structural Steel – Proceedings of 4th Pacific Structural Steel Conf., Vol. 1, Steel Structures, pp. 119-126.

Maleck (Surovek), A.E. and White, D.W. (1998). "Effects of Imperfections on Steel Framing Systems", *Proceedings - 1998 Annual Technical Session and Meeting, Structural Stability Research Council*, pp. 43-52.

Maleck (Surovek), A.E. (2001). *Second-Order Inelastic and Modified Elastic Analysis and Design Evaluation of Planar Steel Frames*, Ph.D. Dissertation, Georgia Institute of Technology, 579 pp.

Martinez-Garcia, J.M. (2002), *Benchmark Studies to Evaluate New Provisions fro Frame Stability using Second-Order Analysis*, MS Thesis, Bucknell University, December.

Martinez-Garcia, J.M., and Ziemian, R.D. (2006). "Benchmark Studies to Compare Frame Stability Provisions", *Proceedings – 2006 Annual Technical Session and Meeting, Structural Stability Research Council*, San Antonio, TX (8-11/2), pp. 425-442

McGuire, W. (1995a). "Inelastic Analysis and Design of Steel Frames. A Case in Point", Proceedings *of Conference on Research Transformed into Practice: Implementation of NSF Research*, Arlington, VA, p 576.

McGuire, W. (1995b). "Inelastic Analysis and Design in Steel, A Critique", *Restructuring America and Beyond, Proceedings of Structures Congress XIII*, M. Sanayei (ed.), ASCE, pp. 1829-1832.

McGuire, R., Gallagher, R., and Ziemian, R., (2000), *Matrix Structural Analysis 2nd Edition*, John Wiley and Sons, New York.

Rex, C.O. and Goverdhan, A.V. (1998). "Consideration of Leaner Columns in PR Frame Design", *Frames with Partially Restrained Connections,* edited by J. Ricles, R. Bjorhovde and N. Iwankiw, Workshop Proceedings, Structural Stability Research Council, Atlanta, GA. 1998.

Rex, C.O. and Goverdhan, A.V. (2002). "Design and Behavior of a Real PR Building", *Connections in Steel Structures IV: Behavior, Strength and Design*, edited by R. Leon and W. S. Easterling, Proceedings of the Fourth Workshop on Connections in Steel Structures, Roanoke, VA. 2002.

Roeder, C., Coons, R.G. and Hoit, M. (2000), *Simplified Design Models for Predicting the Seismic Performance of Steel Moment Frame Connections*, Report No. SAC/BD-00/15, SAC Steel Project Background Document, SAC Joint Venture.

Schneider, S.P. and Teerapabwong, I. (2000), Bolted Flange Plate Connections, Report No. SAC/BD-00/05, SAC Steel Project Background Document, SAC Joint Venture.

Swanson, J.A. Leon, R.T. (2001), "Stiffness modeling of bolted T-stub connection components," *Journal of Structural Engineering*, v 127, n 5, p 498-505.

Surovek-Maleck, A.E. and White, D.W. (2003). "Direct Analysis Approach for the Assessment of Frame Stability: Verification Studies," *Proceedings - Annual Technical Session and Meeting, Structural Stability Research Council,* Baltimore, April, pp. 423-441.

Surovek-Maleck, A.E. and White, D.W., (2004). Alternative Approaches for Elastic Analysis and Design of Steel Frames. I: Overview, *Journal of Structural Engineering*, ASCE, Vol. 130, No. 8, August, pp. 1186-1196.

Surovek, A.E. and Ziemian, R.D. (2005). "The Direct Analysis Method: Bridging the Gap from Linear Elastic Analysis to Advanced Analysis in Steel Frame Design," *Proceedings of the 2005 Structures Congress and Exposition, Metropolis and Beyond*, New York, p 1197-1210.

Trahair, N.S. and Chan, S. L. (2003). "Out-of-Plane Advanced Analysis of Steel Structures, Engineering Structures, V. 25, No. 13, pp. 1627-1637.

White, D. W. (1993), "Plastic-Hinge Methods for Advanced Analysis of Steel Frames," *Journal of Constructional Steel Research*, Vol. 24 No. 2, p. 121-152

Vogel, U. (1985) "Calibrating Frames," *Stahlbau*, Vol. 54, pp. 295-301.

White, D.W. and Chen, W.F., eds. (1993). *Plastic Hinge Based Methods for Advanced Analysis and design of Steel Frames – an assessment of the State of the Art*, Structural Stability research Council, Bethlehem, PA, 299 pp.

White, D.W. and Nukala, P.K.V.N. (1997). "Recent Advances in Methods for Inelastic Frames Analysis: Implications for Design and a Look Toward the Future," *Proceedings, National Steel Construction Conference, American Institute of Steel Construction*, pp. 43-1 to 43-24.

White, D.W., Surovek, A.E., Alemdar, B.N., Chang, C.J., Kim, Y.D., and Kuchenbecker, G.H. (2006). "Stability Analysis and Design of Steel Building Frames Using the 2005 AISC Specification," *International Journal of Steel Structures*, KSSC, 71-91.

White, D.W. and Griffis, L.G. (2007). "Stability Design of Steel Buildings: Highlights of a New AISC Design Guide," *Proceedings, North American Steel Construction Conference*, New Orleans, LA.

Ziemian, R.D. (1990). *Advanced Methods of Inelastic Analysis for in the Limit States Design of Steel Structures*, Ph.D. Thesis, Cornell University, Ithaca, N.Y.

Ziemian, R.D. and McGuire, W. (1992). "A Method for Incorporating Live Load Reduction Provisions in Frame Analysis," *Engineering Journal*, AISC, Vol. 29, No.1, pp 1- 3.

Ziemian, R.D, McGuire, W., and Deierlein, G. (1992), "Inelastic Limit States Design. Part I: Planar Frame Studies", *Journal of Structural Engineering*, v 118, n 9, Sep, 1992, p 2532-2549

Ziemian, R.D., (1993). "Examples of Frame Studies Used to Verify Advanced Methods of Inelastic Analysis," in White, Donald W. and Chen, W. F., eds. (1993), *Plastic Hinge Based Methods for Advanced Analysis and design of Steel Frames – an assessment of the State of the Art*, Structural Stability research Council, Bethlehem, PA, pp. 217-244.

Ziemian, R.D. and McGuire, W. (2002), "Modified Tangent Modulus Approach, a Contribution to Plastic Hinge Analysis," *Journal of Structural Engineering*, ASCE, v 128, n 10, October, 2002, p 1301-1307

Ziemian. (ed.) (2010). *Guide to Stability Design Criteria for Metal Structures*, 6th Edition, Structural Stability Research Council, Wiley, 1078 pp.

Index